Amazing Animals
Gorillas

Michael De Medeiros

WEIGL PUBLISHERS INC.

Published by Weigl Publishers Inc.
350 5th Avenue, Suite 3304
New York, NY USA 10118-0069

Library of Congress Cataloging-in-Publication Data

De Medeiros, Michael.
Gorillas / Michael De Medeiros.
p. cm. – (Amazing animals series)
ISBN 1-59036-390-6 (hard cover : alk. paper)
– ISBN 1-59036-396-5 (soft cover : alk. paper)
1. Gorilla–Juvenile literature. I. Title. II. Series.
QL737.P96D4 2006
599.884–dc22

2005027268

Printed in the United States of America
1 2 3 4 5 6 7 8 9 0 10 09 08 07 06

COVER: Gorillas use body language, facial expressions, and sounds to communicate. They also have feelings. Gorillas laugh when they are tickled.

Editor
Heather C. Hudak
Design and Layout
Terry Paulhus

About This Book

This book tells you all about gorillas. Find out where they live and what they eat. Discover how you can help to protect them. You can also read about them in myths and legends from around the world.

Words in **bold** are explained in the Words to Know section at the back of the book.

Useful Websites

Addresses in this book take you to the home pages of websites that have information about gorillas.

All of the Internet URLs given in the book were valid at the time of publication. However, due to the dynamic nature of the Internet, some addresses may have changed, or sites may have ceased to exist since publication. While the author and publisher regret any inconvenience this may cause readers, no responsibility for any such changes can be accepted by either the author or the publisher.

Contents

Meet the Gorillas

Gorillas are big, strong, intelligent animals. They live in the wet forests of West and Central Africa.

Most people think that gorillas are scary animals. Gorillas are very calm, peaceful, and kind. These big, fearsome-looking animals are **herbivores**.

Gorillas are the largest members of the great ape family. Gorillas also belong to a group of animals called **mammals**.

▼ Gorillas are a mixture of strength and gentleness.

▲ Each gorilla has its own unique facial features.

The Great Ape Family

- The great ape family includes gorillas, chimpanzees, orangutans, and bonobos.
- There are three different kinds of gorillas. They are the western lowland gorilla, the eastern lowland gorilla, and the mountain gorilla.

A Very Special Animal

Gorillas travel in large groups called troops or bands. Troops can have 6 to 30 gorillas. The leader of the group is the **silverback** male. The silverback male is the oldest male in the group.

Like humans, gorillas have an **opposable thumb** and toe. They use them to hold objects. Unlike humans, gorillas are quadrupeds. This means that they use their long arms and legs to walk. Walking on all four limbs is called **knuckle-walking.**

▼ Scientists can identify a gorilla by its nose.

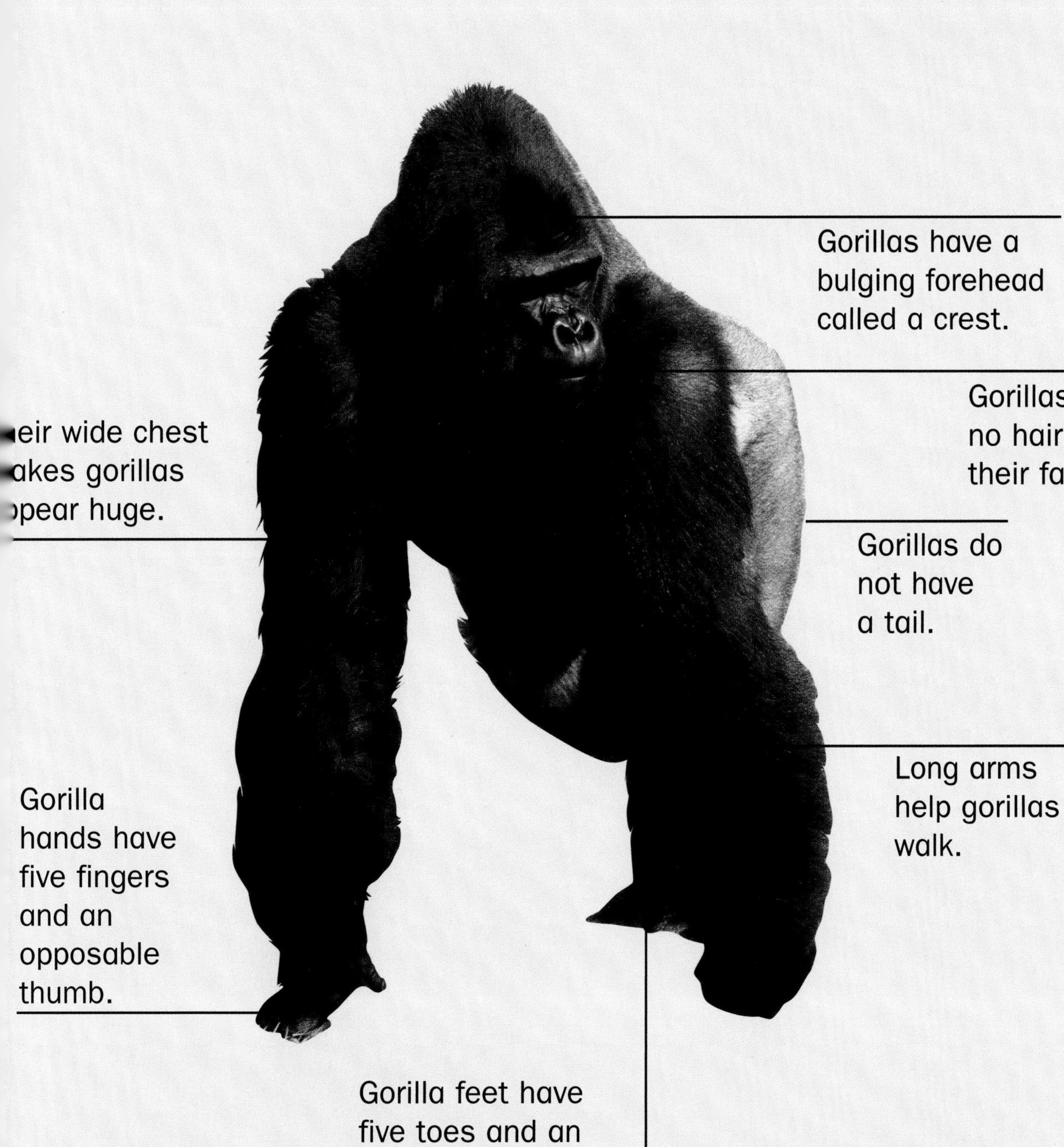

Gorillas have a bulging forehead called a crest.
Gorillas have no hair on their face.
ıeir wide chest ıakes gorillas ıpear huge.
Gorillas do not have a tail.
Long arms help gorillas walk.
Gorilla hands have five fingers and an opposable thumb.
Gorilla feet have five toes and an opposable toe.

Chest Beating

Gorillas beat their chest when they are excited, angry, or frightened. They also stand on their hind legs and make hooting noises. Chest beating is used to communicate with other animals.

Gorillas beat their chest by cupping their hands, making a loud noise. After beating their chest, they tear up all the plants and ground around them. Gorillas will then thump the ground with their open palms.

▶ Gorillas can make 15 different sounds.

Gorilla Talk

- Gorillas show their big, sharp teeth and make loud noises when they are angry or protecting their **territory**.
- Chest beating can also be used as a way of showing off or relieving stress.

How Gorillas Eat

Gorillas mostly eat plants. They will also eat the leaves and stems of different plants. The plants they eat include bamboo, fruit, flowers, and wild celery.

Gorillas do not drink water. They get the water they need from the leaves in their diet.

Gorillas eat sitting down, and their stomachs stick out between their legs. This position makes them look like they have potbellies.

▼ Gorillas eat a large amount of plants to get enough energy for their large bodies.

What a Meal!

- Gorillas eat up to 50 pounds (23 kilograms) of food each day.
- Gorillas spend 6 or 7 hours eating every day.
- Gorillas have **microorganisms** inside their stomachs. Microorganisms help them digest the large amounts of leafy foods they eat every day.

▲ In the forest, there are plenty of lush, green plants for gorillas to eat year-round.

Where Gorillas Live

Gorillas are found only in the forests of Central and Western Africa. They live in thick bamboo forests, warm tropical rain forests, and evergreen mountain forests. Some gorillas also live in city zoos.

Gorillas build two nests every day. Adult gorillas build their nests to rest and sleep in.

▶ Gorillas are found along forest edges and grasslands where there are plenty of plants to eat.

Gorilla Range

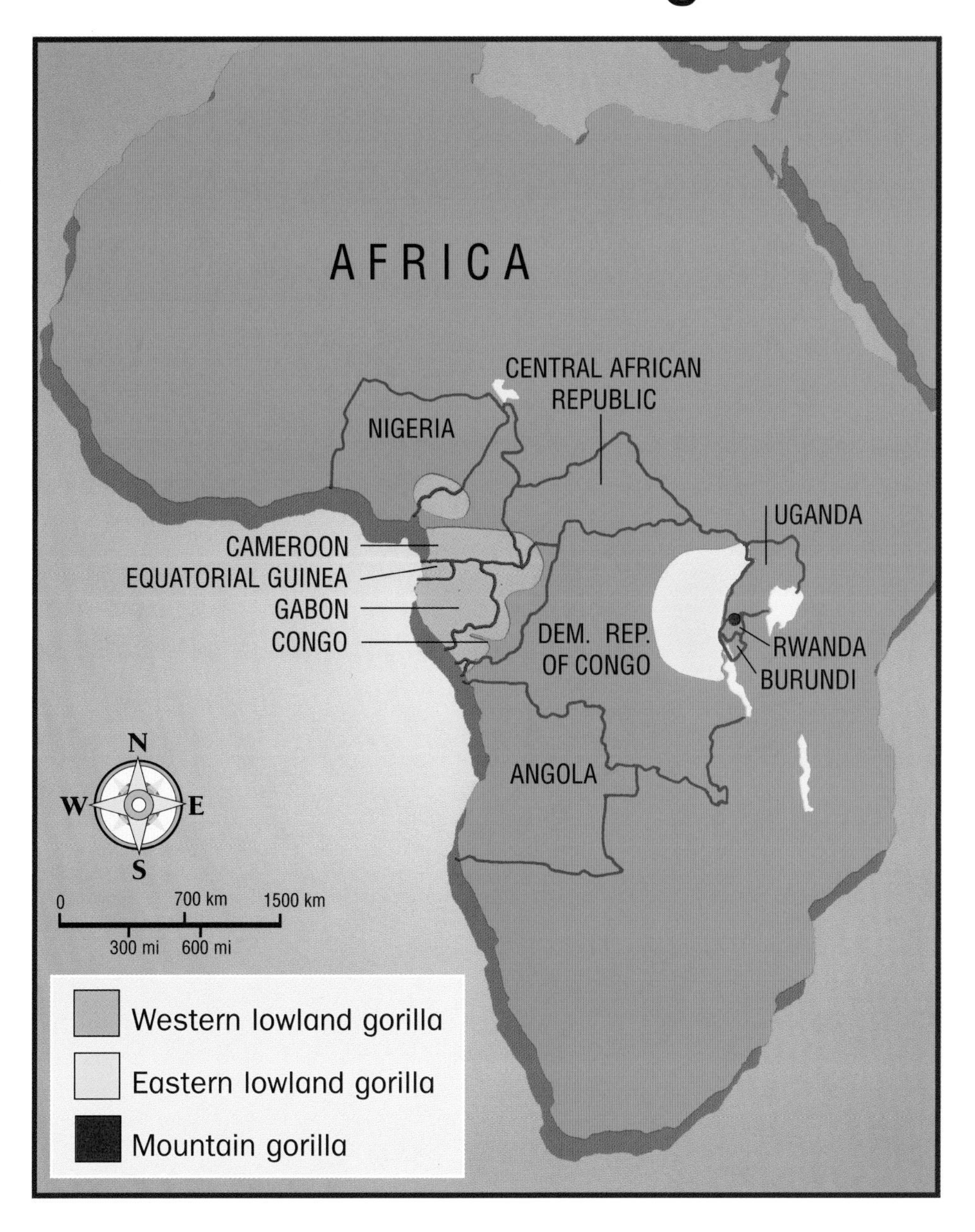

Friends and Enemies

Gorillas peacefully share the forest with other animals, such as antelopes, buffaloes, and elephants.

Gorillas do not have many natural enemies. Leopards are known to attack them. Gorillas also face threats from other gorillas and humans. Humans are by far the biggest threat to them. Humans hunt gorillas for food, sport, and trophies.

▼ Gorillas are very social within their group.

Useful Websites

www.greenpeace.org

Search African forests at this website to learn more about the other animals gorillas live with.

▲ Leopards are found in parts of south and northeast Africa, where they share their homes with gorillas and many other animals.

Living with Gorillas

There are many different animals that live in the same places as gorillas.

- antelopes
- buffaloes
- chimpanzees
- Congo peacocks
- elephants
- monkeys

Growing Up

Gorilla infants are born with pink skin and no hair. They weigh 3 to 4 pounds (1.4 to 1.8 kg). Infants depend on their mothers for food and protection. At birth, gorilla infants are alert and curious.

The mother is very protective of her infant. She carries her infant everywhere she goes for three months.

Young gorillas, called juveniles, help form social bonds within the group.

▶ By the age of two, gorilla infants are playful and independent.

Growth Chart

Birth	3 to 4 pounds (1.4 to 1.8 kg)	An infant is small, helpless, and pink at birth.
1 Year Old	15 to 20 pounds (6.8 to 9 kg)	The skin on their bodies turns black.
3 and half years old	60 pounds (27 kg)	Young gorillas learn to climb trees.
3 to 8 years old	120 pounds (54 kg)	Young gorillas are known as juveniles.
8 to 12 years old	Females: 150 to 250 pounds (68 to 113 kg)	Female gorillas are able to have babies.
15 years old	Males: 300 to 400 pounds (135 to 180 kg)	Silver hair appears on the male's back, and he starts trying to form his own group.

▲ By 12 months, gorilla infants can ride on their mother's back.

Under Threat

Today, there are very few gorillas left in the world. Logging, hunting, and war threaten their population.

Gorillas are an **endangered species**. Scientists believe that less than 10 percent of their **habitat** will remain in 25 years. There are fewer than 700 mountain gorillas and 250 **Cross River gorillas** surviving in the wild. There is no official law that protects gorillas.

▼ Gorillas are among the most endangered species in the great ape family.

Useful Websites

www.dianfossey.org

Visit this website to learn about how people are trying to help save gorillas.

▲ Gorillas can no longer wander freely in the forests because of the dangers they face from illegal hunters and habitat loss.

What Do You Think?

Zoos have helped increase the gorilla populations. Gorillas are held in **captivity** in zoos. This experience can affect their sense of family and social bond. There is also very little space in zoos. Should gorillas be held in captivity? Should they be free in their natural habitat?

Myths and Legends

Movies and books have described gorillas as ferocious animals. Gorillas are not scary animals that attack people. Gorillas usually avoid contact with humans.

Some people think that gorillas are lazy, slow, and dull. In fact, they are strong and muscular. Gorillas are smart as well. They can communicate with each other and with other animals. Some gorillas have even learned to communicate with humans by using sign language.

▶ The 1933 movie *King Kong* was about a scary and dangerous 30-foot (9-meter) gorilla that attacked humans.

Bigfoot

Some people believe that the mysterious creature called the yeti is actually a large, gorilla-like animal. In 1960, Sir Edmund Hillary, a New Zealand explorer, began a journey to find the yeti. Hillary discovered giant footprints near the top of Mount Everest. He was unable to prove the yeti existed. In North America, similar creatures have been called bigfoot or sasquatch.

▼ Most people describe bigfoot as a creature that is large, hairy, and able to walk on two feet.

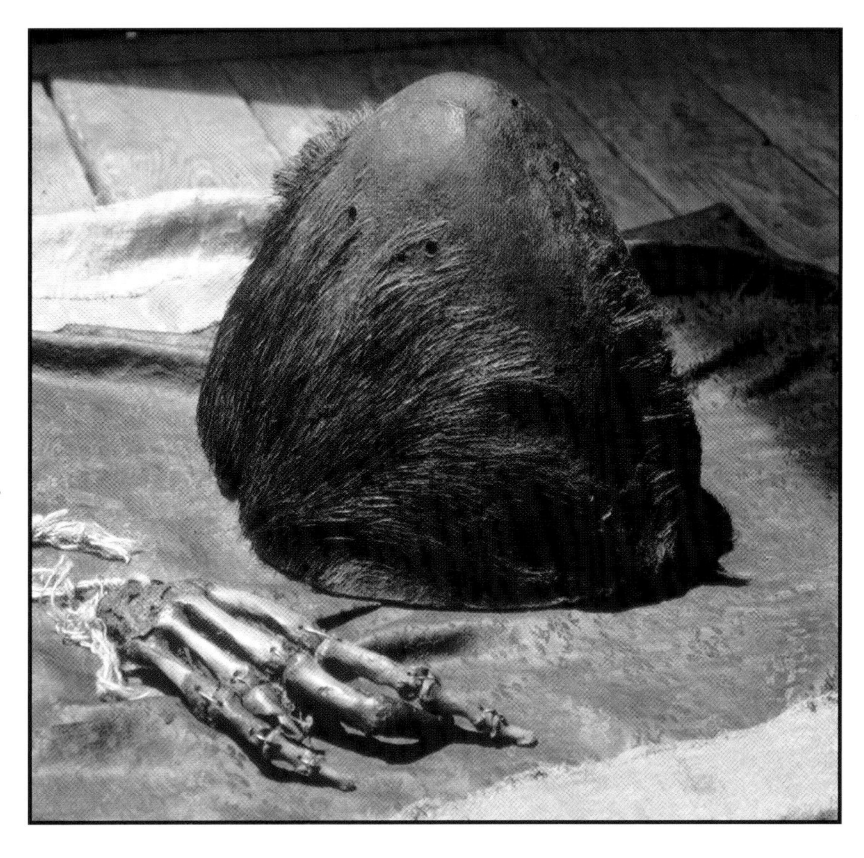

Quiz

1. Where do gorillas live?
   (a) **forests** (b) **farms** (c) **jungle**

2. How many different kinds of gorillas are there?
   (a) **three** (b) **five** (c) **two**

3. What do gorillas eat?
   (a) **meat** (b) **plants** (c) **fish**

4. Who leads the groups, or troops, of gorillas?
   (a) **the oldest female** (b) **the blackback male**
   (c) **the silverback male**

5. What animal has been known to attack gorillas?
   (a) **the leopard** (b) **the elephant** (c) **the antelope**

Answers:
1. (a) Gorillas live in forests.
2. (a) There are three different kinds of gorillas.
3. (b) Gorillas eat plants.
4. (c) The silverback male leads the group.
5. (a) The leopard is the only animal known to hunt gorillas.

Find out More

To find out more about gorillas, visit the websites in this book. You can also write to these organizations.

The Gorilla Foundation
P.O. Box 620-530
Woodside, CA 94062

International Primate Protection League
P.O. Box 766
Summerville, SC 29484

World Wildlife Fund United States
1250 24th Street NW
Washington, DC 20037

Words to Know

captivity
the state of being confined
Cross River gorilla
a certain type of lowland gorilla
endangered species
at risk of no longer living on Earth
habitat
the natural environment in which animals and plants live
herbivores
animals that eat plants
knuckle-walking
to move with feet flat on the ground and weight carried on the backs of fingers
mammals
animals that have hair or fur and feed milk to their young
microorganisms
tiny life forms too small to see with the human eye
opposable thumb
special thumb to help hold objects
silverback
an adult male gorilla with silver hair on its back who leads the group
territory
an area an animal defends against intruders

Index